PROPERTY OF

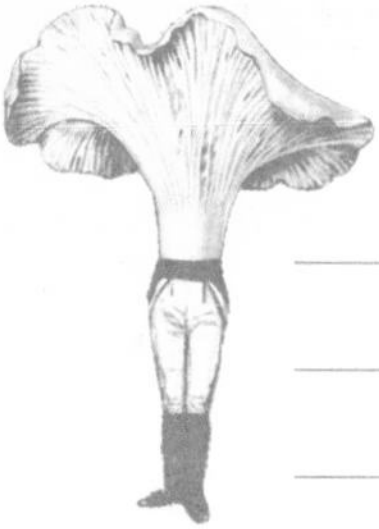

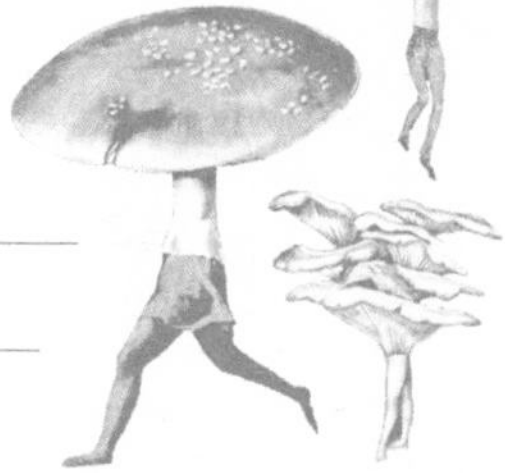

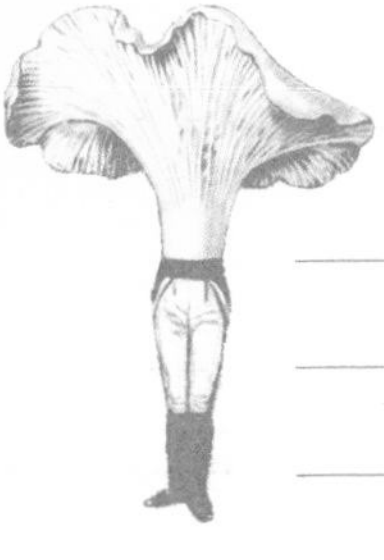

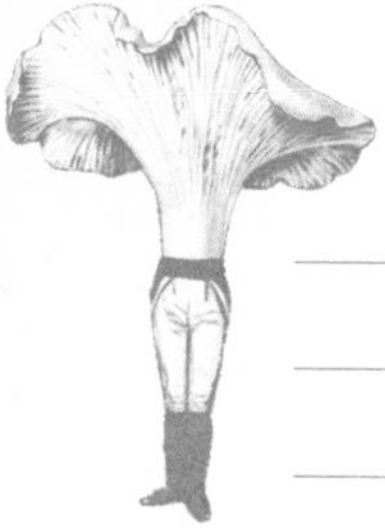

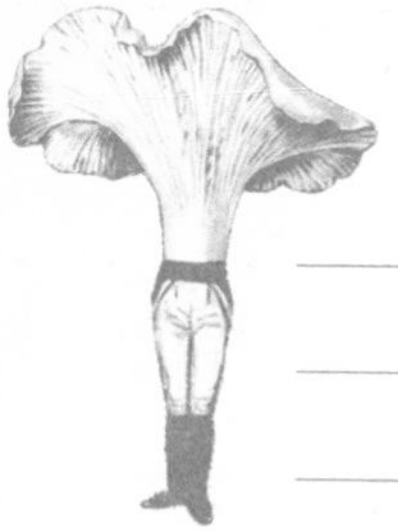

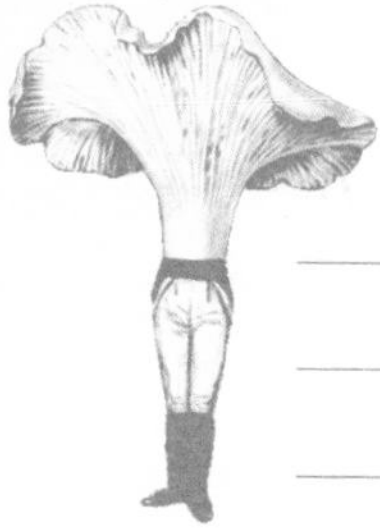

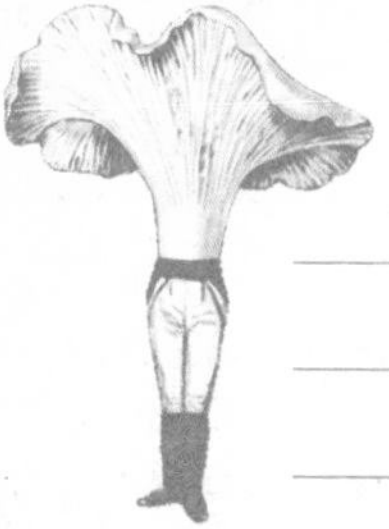

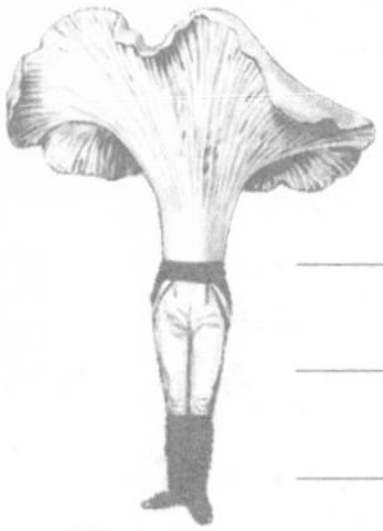

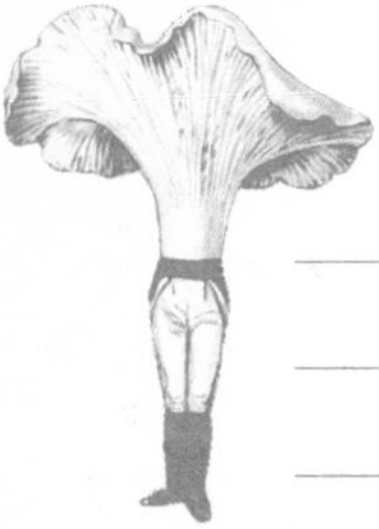

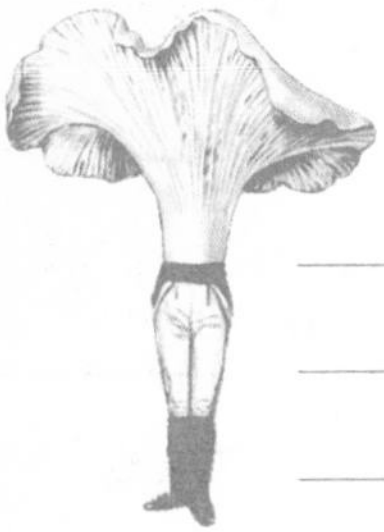

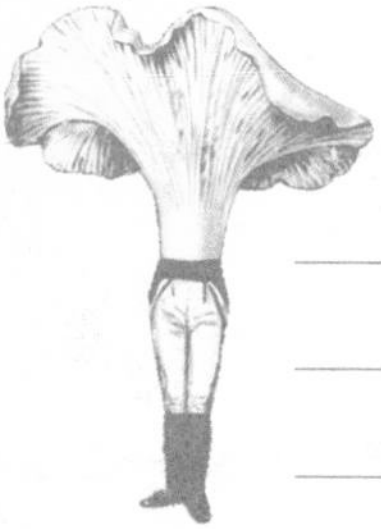

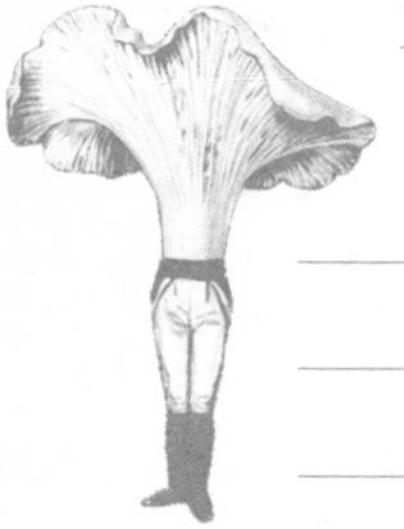

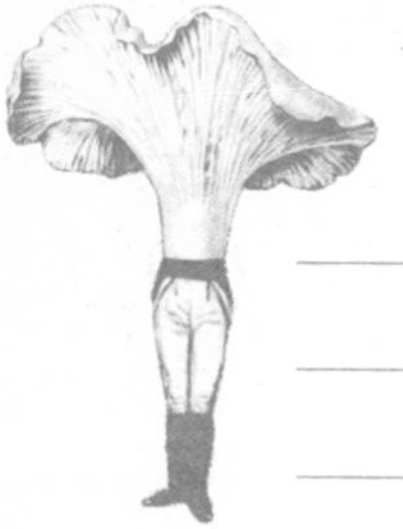

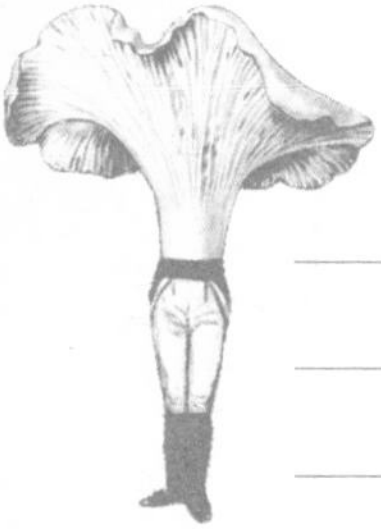

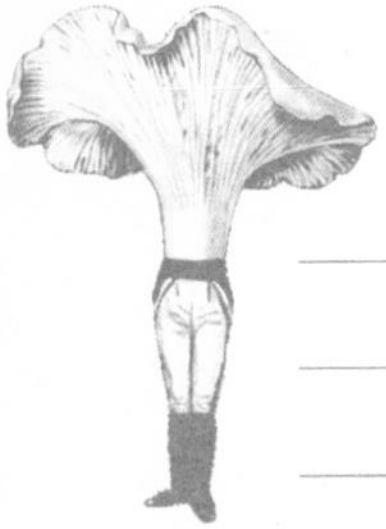

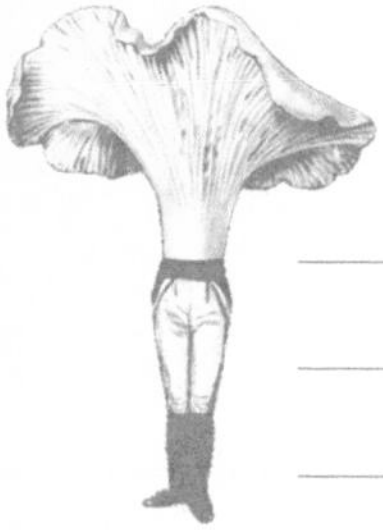

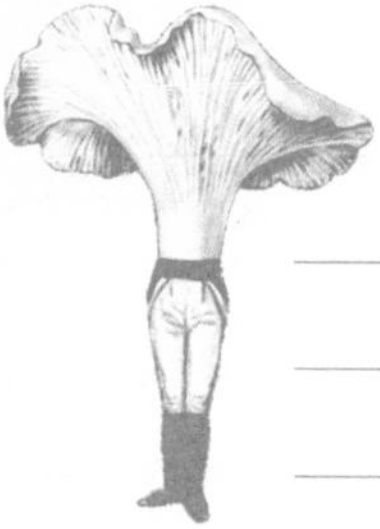

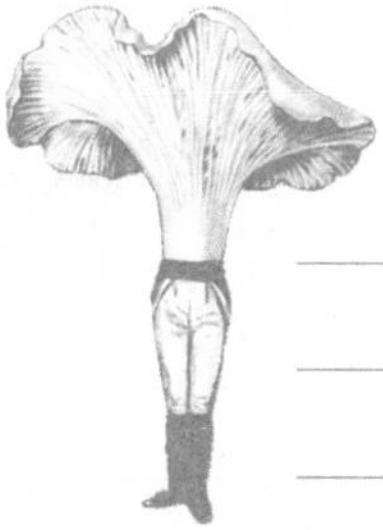

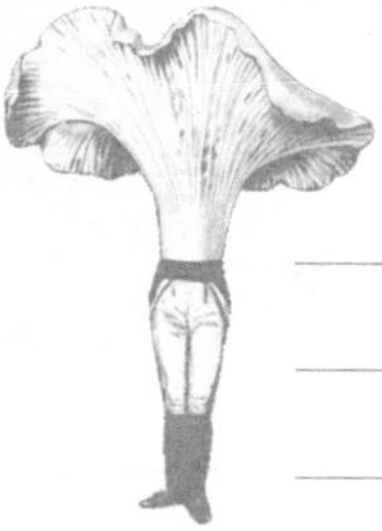

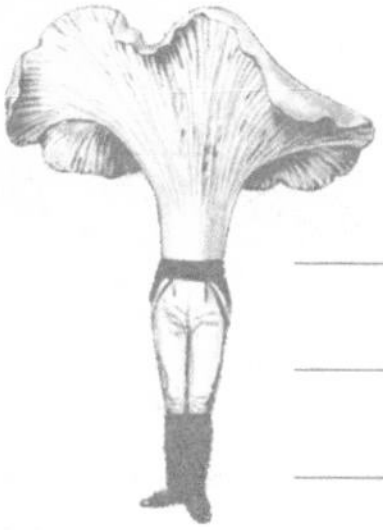

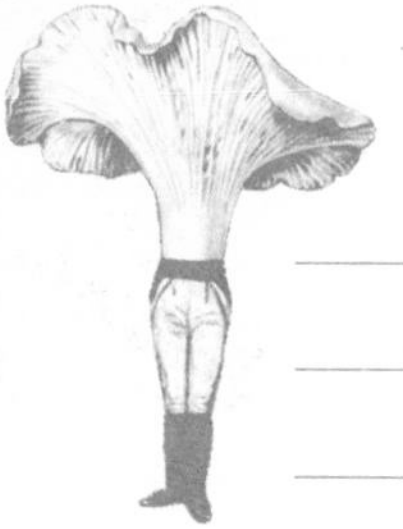

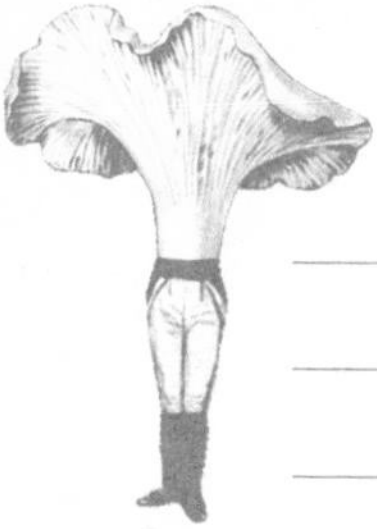

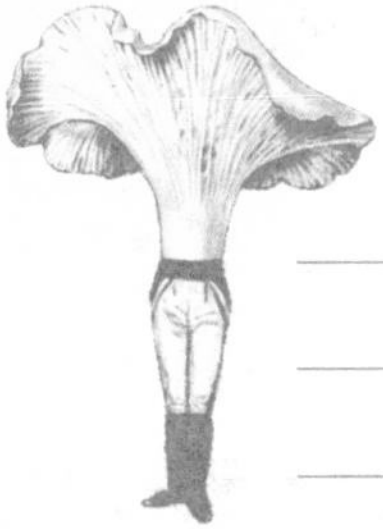

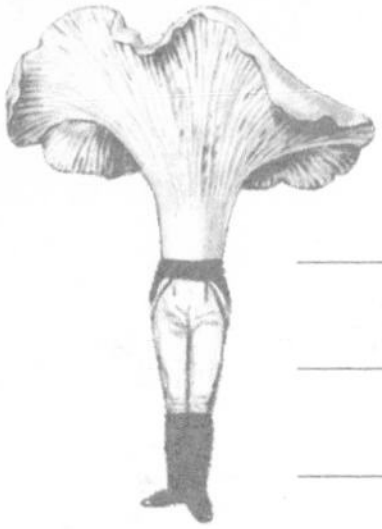

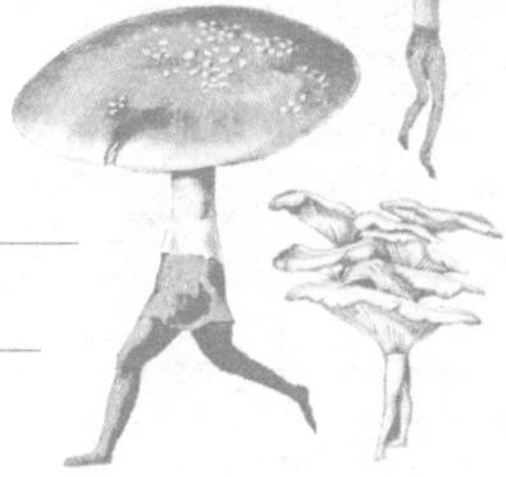

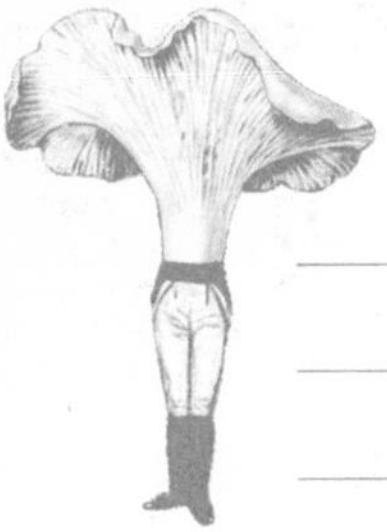

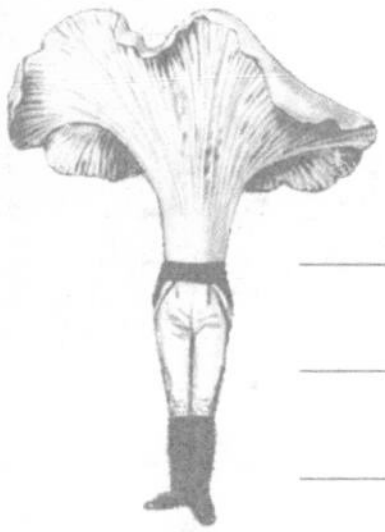

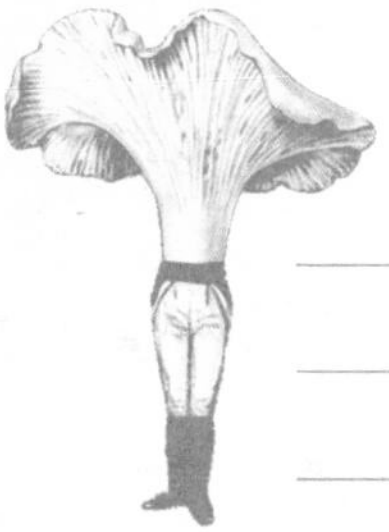

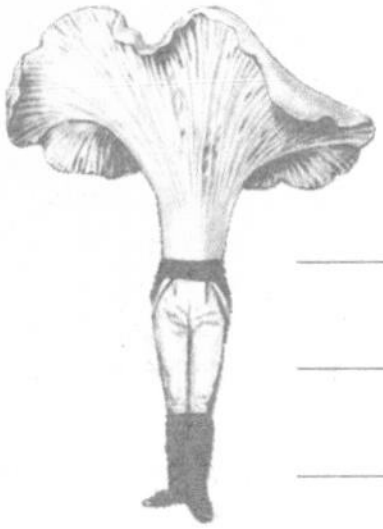

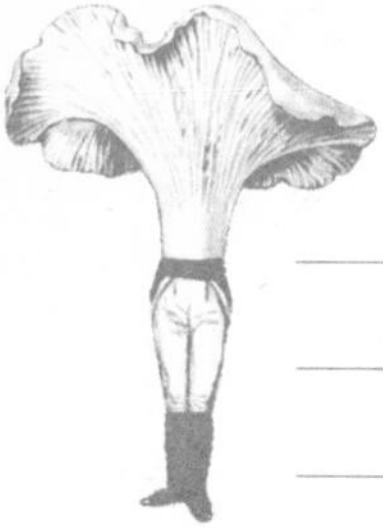

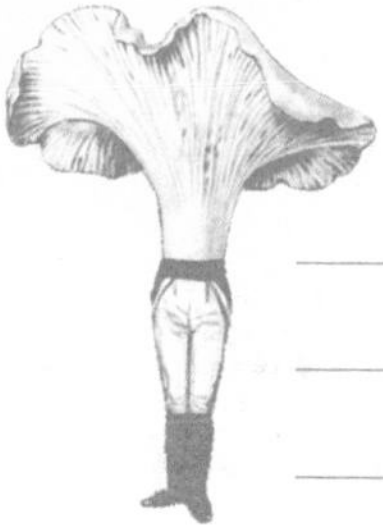

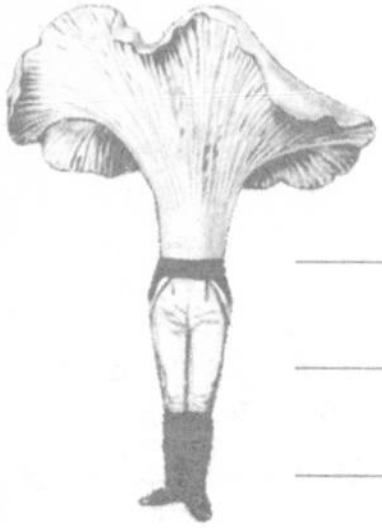

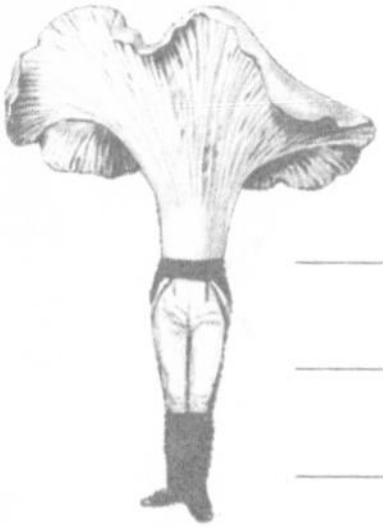

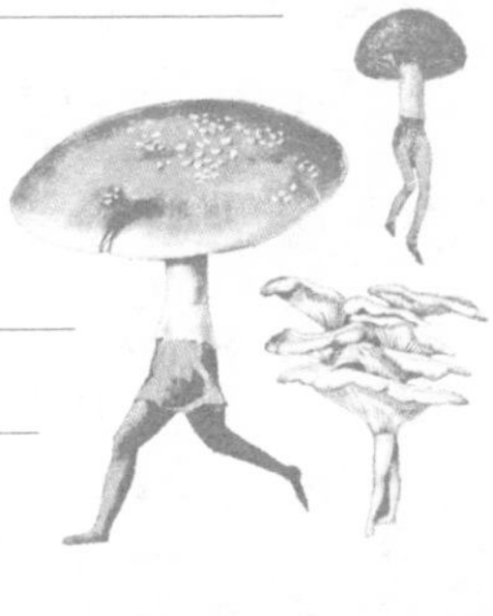

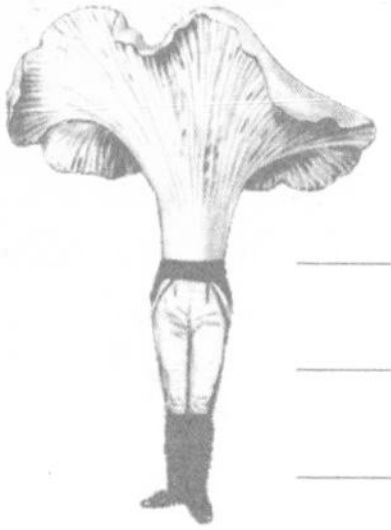

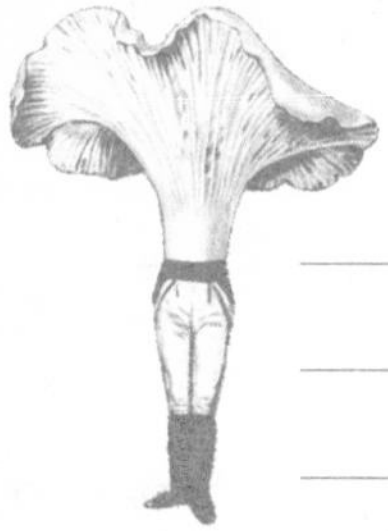

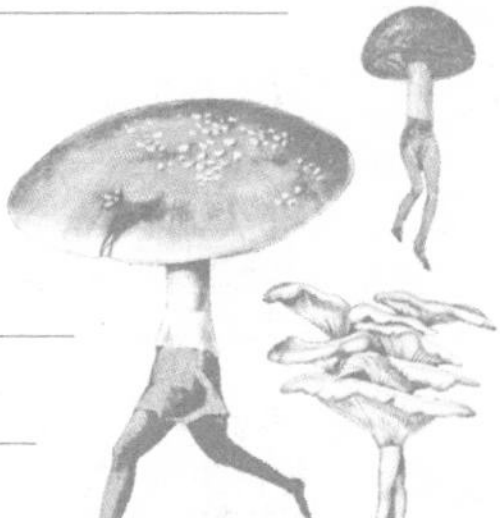

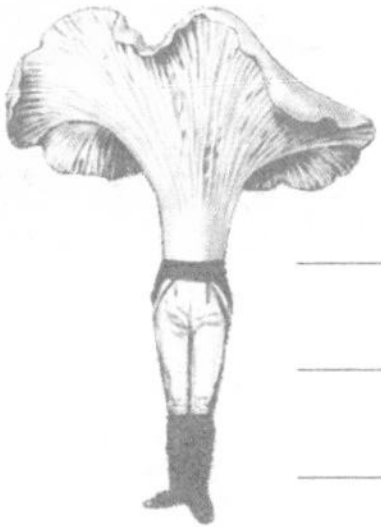

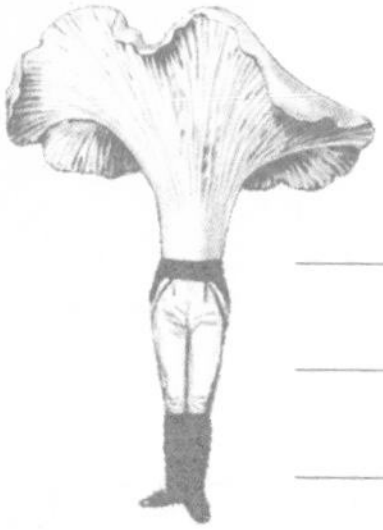

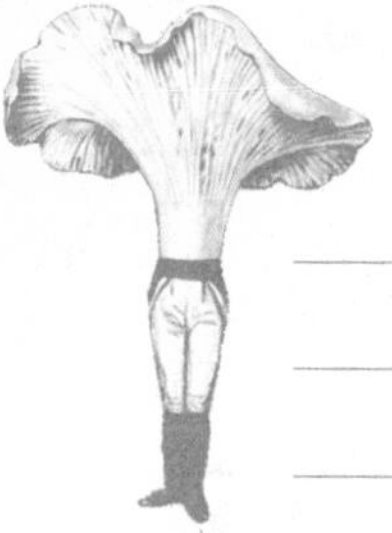

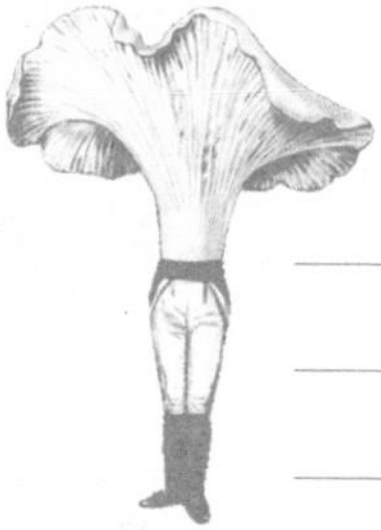

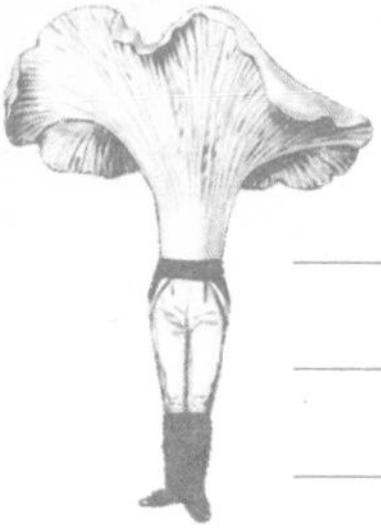

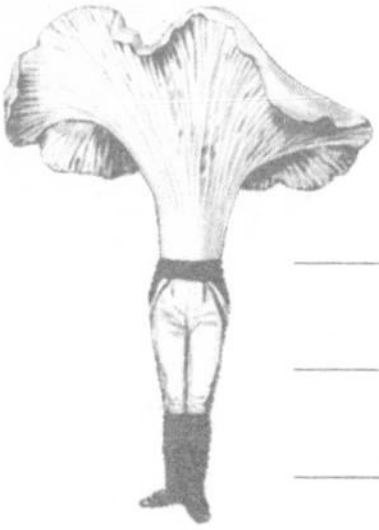

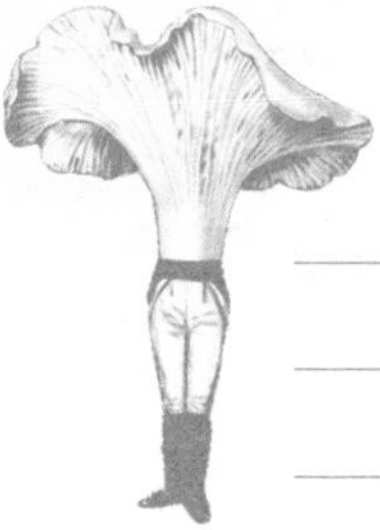

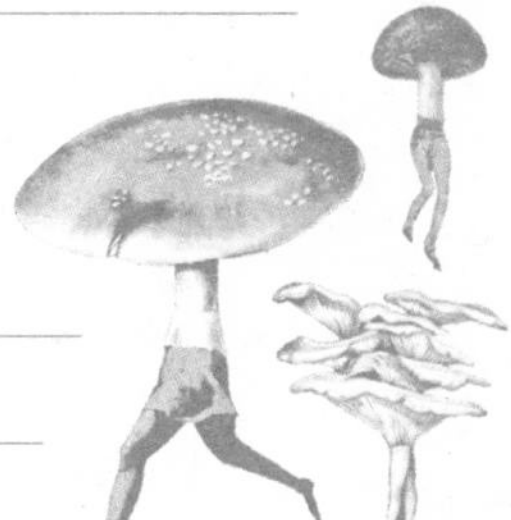

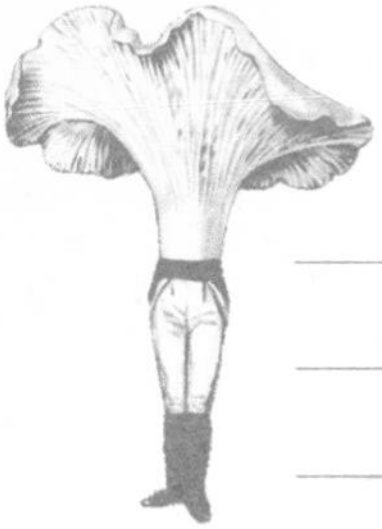

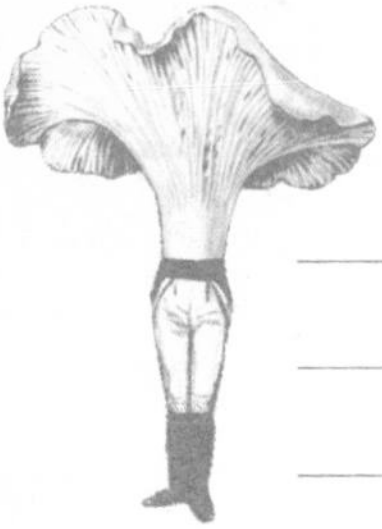

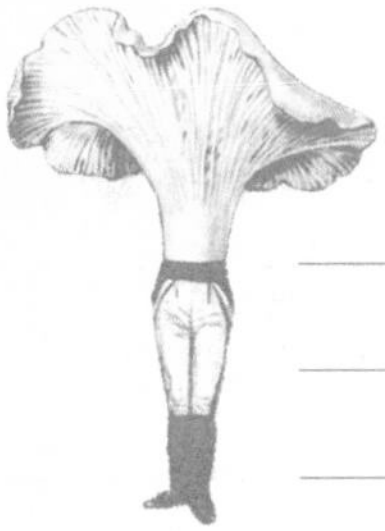

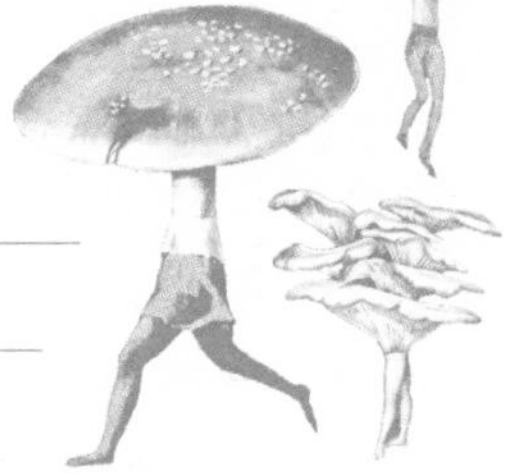

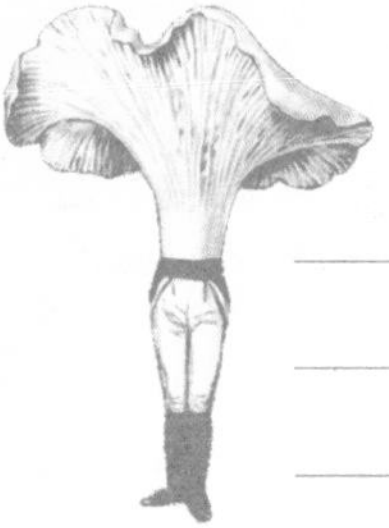

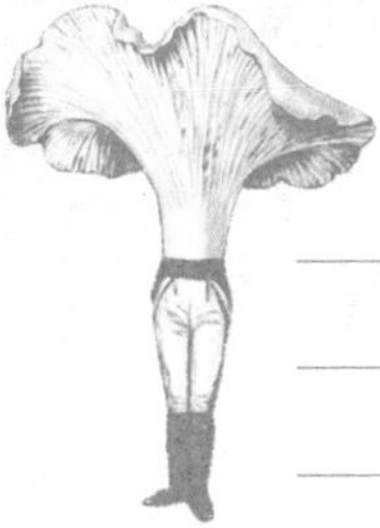

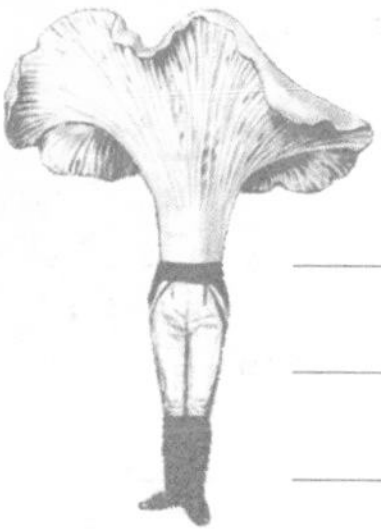

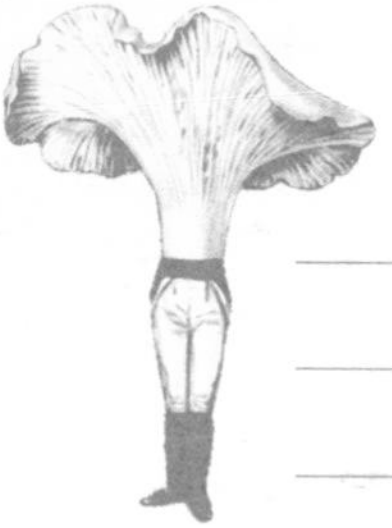

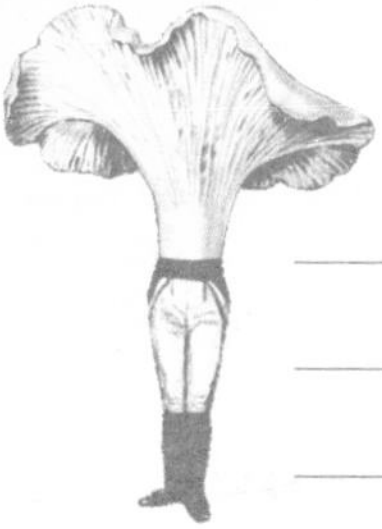

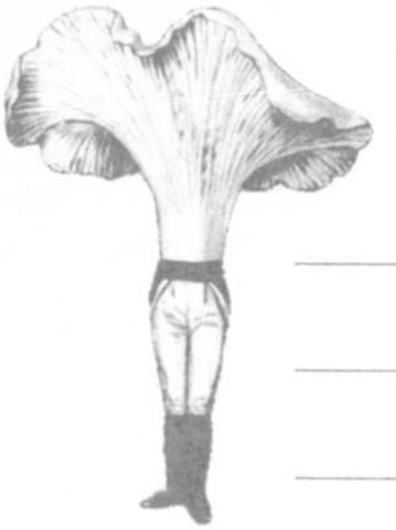

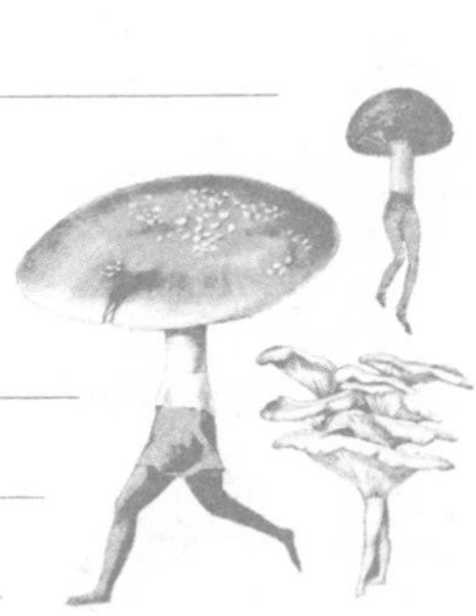

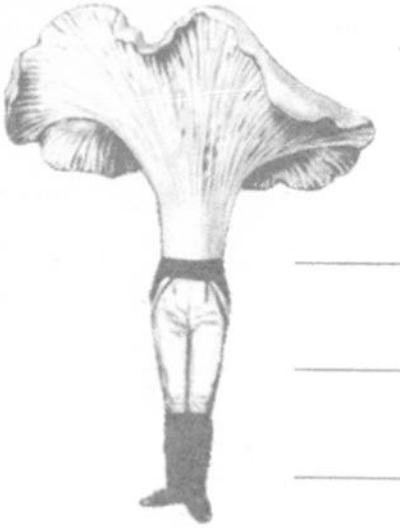

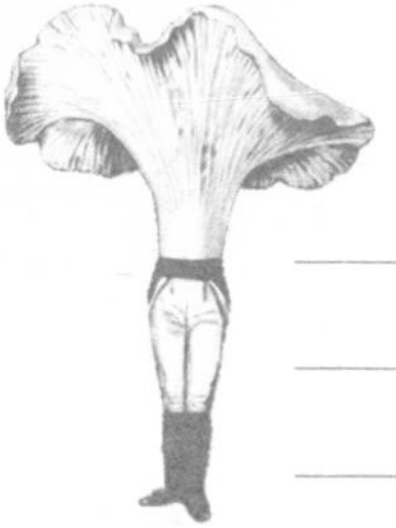

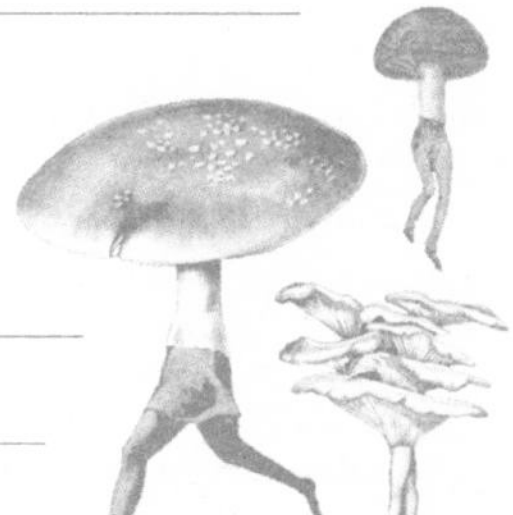

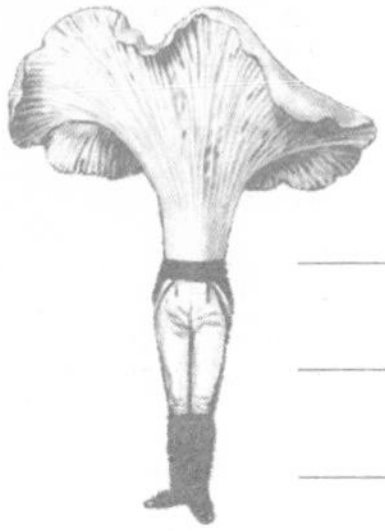

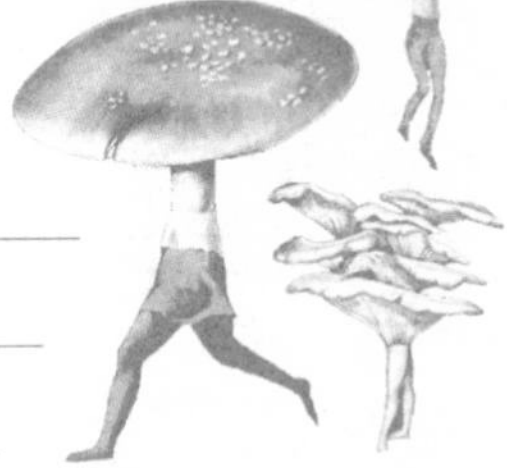

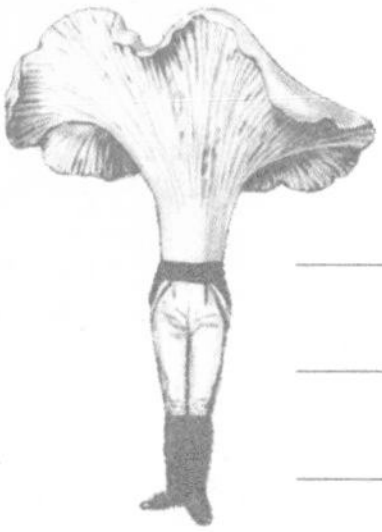

ART OF NATURE
FUNGI COLLECTION

INSIGHTS
an imprint of
INSIGHT EDITIONS
www.insighteditions.com

MANUFACTURED IN CHINA
10 9 8 7 6 5 4 3 2

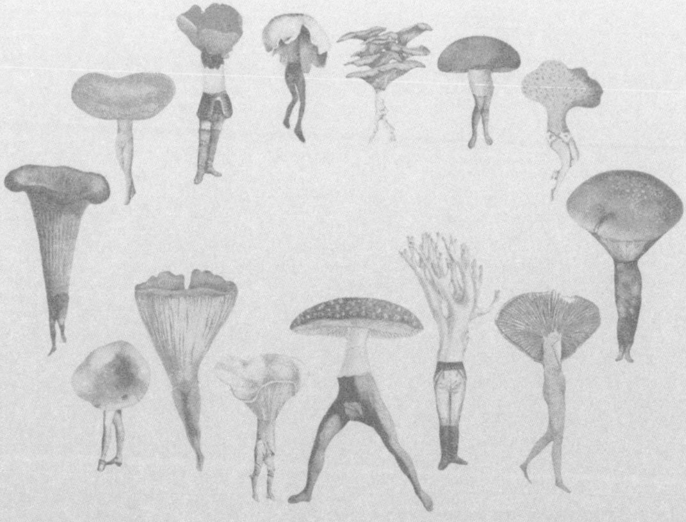